America's New Destiny in Space

On May 24, 2020, my friend and best-selling science fiction author, Steve Stirling, wrote on Facebook:

> *Universal note: in only three days, astronauts will launch from American soil in an American spaceship, for the first time since 2011.*
>
> *And for the first time ever, it will be on a reusable booster.*
>
> *And down in Boca Chica, TX, work proceeds at an incredible pace on multiple iterations of the Starship prototype – SN6 is now under construction, while SN5 gets the finishing touches and SN4 prepares to do its hop test.*
>
> *That's the reusable interplanetary transport that will give us the Solar System the way the Iberian caravels of the 1400s gave us the World Ocean and began the modern world.*
>
> *High-level testing this year, and commercial launches to LEO [Low Earth Orbit] in 2021, if all goes well.*
>
> *That's 150 tons or 100 people to LEO, at $1.5 million/$2 million a pop, and the same amounts to anywhere in the Solar System with orbital refueling, which is part of the package.*

By then SpaceX plans to be building 3 of them a week.

I suspect that in the long term, that will be what 2020 is remembered for; the pandemic will be a footnote.

His statement struck a chord with me. I've been following space developments my entire life. I'm old enough – just – to remember the Gemini and Apollo missions. It's become clear to me that we're in a new and very different phase of space development, one that is likely to be faster, cheaper, and much more sustainable than those that came before. It is also likely to be much more consequential in the long term.

In this short book, I will talk about how we got to this point, what we're doing now, and what's likely to come next. I will also talk about why space exploration matters and which subjects deserve special attention over the coming years.

The Three Phases of Spaceflight

We're entering into a new and different phase of spaceflight, and I think that's an important concept to grasp. Initially, we had what you might call the "visionary" phase of spaceflight. That was followed by the "command-economy" phase (or perhaps we should call it the "steroidal" phase for reasons I'll get to later), which was followed by what now promises to be spaceflight's "sustainable" phase.

In the earliest days of the initial visionary phase, vision was all there was. People were thinking about spaceflight long before they did it. Leaving aside fanciful tales, the first serious thought about spaceflight took place in the late nineteenth century with such works as Jules Verne's novel *From the Earth to the Moon*, which was published in 1867, and Edward Everett Hale's story *The Brick Moon*, published in *Atlantic Monthly* in 1869, which concerned an artificial navigation satellite.

The first serious scientific work (as opposed to fiction) was probably that of a Russian schoolteacher named Konstantin Tsiolkovsky. In 1883, he wrote a monograph about the problems of operating in weightlessness while in orbit. In 1903, he published *Exploration of Cosmic Space with Reactive Devices* which laid out the fundamentals of how to navigate in space using rockets. He continued to explore these subjects for the rest of his life.

Tsiolkovsky envisioned an era in which space

Jeff Bezos and Elon Musk have talked of eventually moving most industry into space.

exploration would lead to cities in space and, ultimately, to utopian societies throughout the solar system. It is a vision shared by many space supporters today. Since the solar system's limitless energy and material wealth would be available to these societies, he reasoned, they would be free from the scarcities that plagued earthbound economies, as well as the conflicts caused by them.

Whether the idea of post-scarcity economics appealed to the Bolsheviks, or whether (more likely) the prospect of rockets as weapons tickled Stalin's fancy, Tsiolkovsky was elevated from being an obscure schoolteacher to a member of the Soviet Academy. There his disciples, such as Sergei Korolev, F. A. Tsander, and Valentin Glushko, began serious work on rocketry that fueled the Soviet Union's pathbreaking space program of the 1950s and 1960s.

Meanwhile, in the United States, pioneer Robert Goddard was experimenting with liquid-fueled rockets. In 1914, he patented the liquid-fueled rocket engine; five years later, he published a paper entitled "A Method of Reaching Extreme Altitudes" that described the prospects for reaching outer space using rockets. Although Goddard was beset by skeptics – including the editors of the *New York Times* who, in 1920, mocked him for considering such far-fetched ideas as sending a rocket to the Moon – he devoted his life to perfecting liquid-fueled rockets. (The *Times* finally retracted its snarky editorial almost fifty years later, after men landed on the Moon in 1969.)

Goddard's work provided inspiration to a number of German rocket pioneers, who began organizing in the early 1920s. They watched as Goddard successfully launched the first liquid-fueled rocket and demonstrated the first working guidance system. Germany's pioneers included Hermann Oberth, who in 1923 published *The Rocket into Planetary Space* (*Die Rakete zu den Planetenaumen*), an ambitious book that looked at the practicalities of putting humans into space in the relative near term, the need for space suits, and the minutiae of operating spaceships and space stations. Oberth's book was followed by numerous others, most notably by Walter Hohmann whose work on celestial mechanics, published in 1925, set out principles still relied on today, and after whom the economical "Hohmann Transfer Orbit" used by deep space probes was named. As the focus shifted to engineering, Wernher von Braun and Willy Ley became involved at the rocket-club level. Rocket engines were designed, tested, and launched (when they didn't explode) and many important technologies, such as the use of fuel to cool the combustion chamber and nozzle, were developed and refined.

In the United Kingdom, the 1875 Explosives Act unwisely prevented private research into ordnance, which effectively and unfortunately barred amateurs from hands-on work with rockets. But the British Interplanetary Society, which included illustrious members such as Phil Cleator and Arthur C. Clarke, turned its efforts to mission design and in 1939 produced a famous plan for a Moon mission that served as the foundation for the actual Apollo Moon landing thirty years later.

But there was only so far you could go as a visionary. In Walter McDougall's *... the Heavens and the Earth: A Political History of the Space Age*, engineer Willy Ley recalled that by the 1930s von Braun's amateur rocketry club had reached the point where continuation would have been "too expensive for any organization except a millionaire's club." Or a government, as it turned out. This is where the visionary phase ended and the command-economy phase began.

As early as 1929, the German government became interested in rockets with ranges greater than any existing cannon. This was a very ambitious goal given that the "Big Bertha" used in World War I had a range of 65 miles. Von Braun and his amateurs soon came to the attention of the German army's Ordnance Ballistic Section and, with funding from the government, they established a test facility and range at Peenemunde on the northern Baltic coast.

When a SpaceX rocket explodes, it's not a political scandal but a learning event for engineers.

By 1942, they were successfully flying the A-4 rocket that, with the addition of a warhead, became the V-2 missile. As the Luftwaffe's fortunes declined, von Braun's missiles became more important, and he and his team refined the technology while working on designs for manned rockets and missiles with intercontinental range. In truth, as Ley had commented, they made progress at a rate that never could have been equaled by a private group.

After the Second World War ended, von Braun and many of his team fled to the West and were scooped up by the United States; some of their less fortunate colleagues were corralled by the Soviets. This was also the beginning of the era when the United States began to see the Soviet Union as its future adversary. As an open society struggling against a closed one, the U.S. was particularly interested in spy satellites (in 1946, a Douglas Aircraft Company/RAND paper went into detail as to the possibilities) as well as potentially missiles, though the U. S. Air Force thought its new jet bombers would be enough. Lacking such an air force of its own, the Soviet Union pursued missile development headlong while the United States lagged behind.

This changed with Sputnik. Without warning, the Soviet Union shocked the world by launching the first-ever orbital satellite in 1957. Most people had no idea that such a thing could be done, much less that the Russians, who were generally viewed as backward and primitive, were capable to doing so. In addition, it was clear that a nation that could put a satellite in orbit could also deliver a nuclear warhead by missile.

This development greatly upset American voters, and soon the United States and the Soviet Union were locked in a "space race." Partly this was a race for prestige and partly it was a displacement of the Cold War conflict. (Jerome Wiesner, a science advisor to

John F. Kennedy, once told me that Kennedy explicitly saw the space race as a way of diverting and diffusing earthbound Cold War tensions into a nonviolent arena.) But there was more. Hanging overhead was the Moon and, based on colonial precedents, it seemed possible that the first nation to reach the Moon could claim it for itself. The prospect of Soviet missile bases orbiting overhead (or, for the Soviets, of American missile bases doing so) was deeply disturbing and both nations set out to get there first.

The result was a much bigger version of von Braun's World War II efforts, fueled by substantially more money. Once again, enormous progress was made in record time. In less than a decade, the United States went from being unable to launch a human being into space to landing multiple human missions on the Moon, and returning the crews alive. We went from repurposed Redstone missiles as boosters to the gigantic custom-designed Saturn V. There were tremendous advances in computing, materials, and project management. We even got a space station (the SkyLab) out of it.

Ultimately, the Soviets were unable to keep up. They built rockets and acquired substantial expertise, but never got close to matching the long-range, human-crewed capabilities of Apollo. They just couldn't afford it.

And, really, neither could the United States. Or rather, having ushered in the "Great Society" with its collection of expensive social programs, the United States was no longer willing to bankroll the space program. The Apollo program, which was never meant to be economically self-sustaining, didn't survive long into the 1970s, though we did manage a few missions (SkyLab in 1973/74 and Apollo-Soyuz in 1975) with the hardware that was still lying around.

The whole endeavor had lost its urgency with the signing of the 1967 Outer Space Treaty, which among other things banned "national appropriation" and fortification of the Moon and other celestial bodies. Both the U.S. and the U.S.S.R. were, in truth, more afraid of the other getting to the Moon first than they were eager to do so themselves. As a result, they agreed to take lunar missile bases and territorial claims off the table. Perhaps not coincidentally, a graph of U.S. space spending shows a peak in 1967, followed by a post-treaty decline.

By the 1980s, after a long lag, Apollo was eventually replaced by the Space Shuttle. But the Space Shuttle, designed as much to keep NASA jobs in key congressional districts as anything else, wasn't economically sustainable either. Sold as a way of making spaceflight routine and inexpensive, it failed at both, winding up fragile and more costly. The entire Apollo/Space Shuttle space program was a species of command economy and though command economies have their virtues, they don't hold up well over time. (A command economy is sort of like using steroids for a bodybuilder – everybody "oohs" and "aahs" over the visible muscles that pop up, but nobody sees the shrinking national testicles underneath.) You can sustain a command economy only as long as there are enough resources coming from elsewhere to keep it going, and its very existence makes that less likely over time.

The Apollo/Space Shuttle era ended in 2011 and with it so did the command-economy phase of spaceflight. Now, in the 2020s, we're seeing its replacement – none too soon – by sustainable spaceflight: spaceflight that generates enough economic value to pay its own way.

Each of these phases have played a crucial role. Without the visionary phase, the command-economy phase would never have happened; or, if it had, it would have been struggling for ideas and understanding at a very late date. Additionally, the command-economy phase demonstrated that ideas that once seemed wild and impossible – visionary, even – were in fact entirely possible, if perhaps a bit expensive. The sustainable phase promises to be very different from those previous phases and even more essential to getting us to where we want to go. In short, the sustainable phase will make the visionaries' visions affordable.

Where We Are Now

At present, we are in the early stages of the sustainable phase, which kicked off roughly when the Space Shuttle stopped flying. The transition was gradual, and started to take form in the 1980s when many space supporters concluded that relying on government to get them where they wanted to go was a bad idea.

The first tentative commercial space launch companies started in the 1980s. Some, like Orbital Sciences Corporation, which was founded by three Harvard MBA graduates, went on to success. Most didn't. In 1984, Congress passed the Commercial Space Launch Act which made it possible for people to engage in the business. After the Challenger disaster took the Space Shuttle offline for an extended period, amendments were made to the Act in 1988 that made the law much more favorable to private enterprise. The early launch successes of these companies – even those that went bust – helped persuade Congress to take the industry seri-

We have all the ingredients for a new era in space: falling prices; abundant resources; and eager entrepreneurship.

ously (as did the money that some of them spent on high-priced lawyers and lobbyists).

Through the 1990s and into the new millennium, advocacy groups like the National Space Society (full disclosure: I was in charge of policy and legislation at the NSS starting in 1989 and became executive vice president from 1991–95) worked to advance the position of private-launch companies, encourage friendly government regulation, promote the growth of a strong private-sector presence in space, and fight the "dumping" (sales below cost that would hurt the development of American companies) of Chinese government-launch services.

The thinking was straightforward: a government-based space program would always be subject to political winds and whims. Not being self-supporting, it would always face competition from other programs that gave a more direct payoff to various political interest groups. In addition, even when funded, a government program would tend to dissipate its resources and efforts in bureaucratic payroll padding and empire building. A commercial enterprise, disciplined by shareholders and the bottom line, would have to focus more on producing results and, more crucially, on lowering costs and improving performance – something that NASA had not distinguished itself in achieving.

With a new family and related distractions, I took about a decade off from the scene. When I returned, the change was obvious. The annual International Space Development Conferences, which once had the ambience of a *Star Trek* convention, seemed far more professional. There were still plenty of button-bedecked enthusiasts, but there were also venture capitalists in Brioni suits and entrepreneurs who were actually bending metal and launching things, or working hard at doing so. I ran into a woman who I remembered as a student activist with Students for the Exploration and Development of Space; now she was a professor of astronautics at MIT. Another former student activist was in charge of space matters for Google.

Summing up the changes in an article for *The Atlantic* in 2008, I wrote:

> *JFK's decision to go to the Moon was a tremendous case of wish-fulfillment for space enthusiasts – instead of a slow growth into outer space, we got a massive commitment of federal resources to get us to the Moon faster than anyone had thought possible. Given that kind of boost, it's not surprising that many spent decades trying to reenact that kind of success. The JFK moment, however, hasn't come around again, and isn't likely to repeat itself any time soon. And, as is often the case when a fantasy becomes reality, the aftermath turned out to be a bit of a letdown; the massive commitment of resources brought about by the political excitement dwindled as soon as that political excitement inevitably disappeared. Politics and command-and-control systems got us to the Moon in a decade, but they proved powerless*

to keep us there. Achieving a space-faring civilization incrementally though profit-making enterprise may take longer, but such an approach doesn't depend on the whims of presidents, or on right guesses by NASA. It's an approach that takes patience – a quality that has arrived, as it often does, through a combination of time and disappointment.

Ironically, today's state of affairs is looking like what a lot of classic – that is, pre-Apollo – science fiction predicted: a slow, steady growth of commercial endeavors in space, with ordinary citizens reaching the Moon in the early twenty-first century via the efforts of a wealthy industrialist.

Since I wrote that, we've made enormous strides thanks to the efforts of wealthy industrialists – particularly Elon Musk, but also many others such as Jeff Bezos, as well as the entrepreneurs and Google founders who funded the X-Prize and Lunar X-Prize competitions that have inspired numerous innovators to apply their talents to space. We've also seen enormous results; results that make it possible to do great things in space in the near future, without depending on political whims.

Cutting the Cost

As we'll see later, there are many things that we can do in space, if we can just get there. But getting there is expensive. The largely unheralded revolution of the last decade has been that getting into space has become far cheaper than it used to be, and that it promises to get much cheaper still.

As the science fiction writer Robert Heinlein famously said to his colleague, Jerry Pournelle: "Once you're in Earth orbit, you're halfway to anywhere" in the solar system. That statement is not mere hyperbole but physical truth. The change in velocity (and hence energy required) to get from Earth to orbit is approximately the same as that needed to reach even the most distant planet in the solar system from Earth's orbit. Thus, a change in the cost of getting to orbit is a pretty good first-order measure of how our changes in space capability are going. Here, the news over the past decade has been very good.

It used to cost almost $55,000 to get a kilogram into orbit on the Space Shuttle. To do the same thing today with the Falcon 9, the newest rocket from Elon Musk's SpaceX enterprise, costs approximately $2,700. That's roughly a twenty-fold reduction.

Many things that are too expensive to do at $55,000 per kilogram become doable at $2,700 per kilogram. And SpaceX is not standing still. Its Starship reusable rocket, now under development, is expected to cost a mere $2 million per launch. Elon Musk says its cost per kilogram to orbit will be at least ten times lower than that of the Falcon 9. There are a lot more things that become doable at approximately $270 per kilogram. At these prices, things like space tourism and hotels, lunar mines, and asteroid mining become feasible. At a certain point, prices get low enough to draw in all sorts of new activity – much as when computing power became so affordable that suddenly it started to appear in things like washing machines and kids' toys. I asked a former NASA official who's very familiar with the industry about this and got this reply: "We are just now, with reusables,

Getting into space has become far cheaper than it used to be.

starting to see possible price elasticity due to lower prices. [...] The Commercial Space Transportation Study in 1994 predicted launch prices needed to drop below $400 per pound for price elasticity to become significant. That would be about $1,550 per kilogram in today's dollars. Almost there."

Faster, please! SpaceX isn't the only company lowering costs in this area; it's just the one making the biggest, most public splash. Other companies, ranging from the relatively tiny RocketLab to Jeff Bezos's secretive Blue Origin, are doing the same.

They're lowering costs because doing so is essential to their business. Additionally, since they are commercial rather than political enterprises, they can afford to fail. When a SpaceX test rocket explodes on the stand, that's not a political scandal or a tragedy, but a learning event for engineers to study in order to find out what went wrong and how to stop it from happening again. Instead of congressional hearings, the problem fix is examined at the next test launch. There are no rewards for boosting payrolls, adding levels of management, or opening a facility in a key congressional district. Customers want cheap, reliable launches, and the only way to survive is to give them what they want. The only way to flourish is to give those customers more of what they want than the competition can.

We're already seeing some of the benefits. SpaceX is launching its Starlink broadband, low-latency satellite network, in which thousands of relatively small, advanced satellites (each weighing approximately 260 kg) will orbit the earth simultaneously, allowing people to access the internet from anywhere. It's doubtful that this could have been profitable at costs of more than $50,000 per kilogram. It's likely that it can be at one-twentieth of the cost.

The space resource used by communications satellites is locational. By virtue of being located overhead, they can see and communicate with large swathes of territory at a time. ("Spy" satellites – more politely referred to as "remote-sensing" satellites – also take advantage of this locational resource, and while they were once the sole domain of superpower intelligence agencies, lower costs have made their commercial use so common it's no longer even newsworthy.) But there are other resources available, too.

Space is full of energy. This energy is clean in the sense that you don't need to do anything on Earth to produce it. It's been known for decades that we can power things on Earth by placing large solar arrays in orbit and beaming the power down via microwaves. This idea was originally proposed by Peter Glaser of the Arthur D. Little company back in 1968. The principle is simple enough: large solar arrays are placed in orbit comprised of panels that can be made gossamer-thin since there's no wind and no gravitational pull to resist. The resulting power is concentrated and converted to microwaves, then beamed to a receiving antenna on Earth where it is converted back to electricity. Early experiments at NASA's Jet Propulsion Laboratory's Goldstone facility in California showed that power could be beamed over distances of miles at efficiencies of greater than 84 percent, which compared favorably with earthbound power transmission. Those tests contributed to a three-year joint NASA/U. S. Department of Energy study of space-based solar power that concluded the technology was feasible. More recently, the U.S. Air Force's experimental X-37 spacecraft conducted more up-to-date power-beaming experiments. The results have not been made public, but it's safe to say that twenty-first century electronics are likely to perform even better than the Reagan-era experiments indicated. (Don't worry about death rays – the power density of the beam is low enough that birds could fly through it safely.)

Though our earthbound supply of fossil fuels

seems much higher than we had thought a few years ago, thanks to the "shale revolution," and though clean nuclear power is always an option, the Solar Power Satellite approach is a way of generating electricity that doesn't require anything to be burned or fissioned on Earth. If desired, we could even use the electricity to split water and create clean-burning hydrogen fuel for vehicles. This generates no greenhouse gases since hydrogen's only combustion product is water. Unlike earthbound solar power, the sunlight is unfiltered by clouds or atmosphere and not subject to Earth's nighttime darkness.

At more than $50,000 per kilogram, these orbital power stations are almost certainly too expensive. However, at one-twentieth of that, they start to look good. At the costs Elon Musk is talking about for the Starliner, they look very good indeed.

Nor are the resources of space limited to intangibles like location or energy. Actual physical resources in space are receiving greater attention, both from investors and governments – including, as we will see, the U.S. government. In the longer term, these may turn out to be the most consequential resources of all, since they can be used not only to benefit people on Earth, but also to support – and make much cheaper – far more ambitious activities in space than can be launched from Earth, even at low cost. For example, Solar Power Satellites could be launched from Earth; ideally, they'd be assembled in space and mostly from space resources which are plentiful.

The solar system contains more than 750,000 asteroids measuring at least one kilometer across. Millions of smaller objects are scattered throughout the solar system, mostly in the asteroid belt between Mars and Jupiter, though some are much closer to Earth. Even a comparatively small asteroid (small enough to move, not just mine) is potentially quite valuable, both on Earth and in space.

A 79 ft. wide M-type (metallic) asteroid could hold up to 33,000 tons of extractable metals, including $50 million in platinum alone. A 23 ft. diameter C-type (carbonaceous) asteroid could hold 24,000 gallons of water, useful for generating fuel and oxygen. Given that one gallon of water weighs 8.33 pounds, it can cost tens of thousands of dollars to launch just that one gallon into Earth's orbit. Even at SpaceX's promised low costs, the numbers would need to improve substantially for water launched from Earth to compete with water that's already floating in space.

Larger asteroids could be worth as much as a superpower's GDP. Asteroid 1986 DA is a metallic asteroid made up of iron, nickel, gold, and platinum. Estimates of its value range between $6–$7 trillion. Unlike some smaller asteroids, something that size won't be retrieved anytime soon, but the figure gives some idea of just how much wealth is out there.

Larger asteroids could be worth as much as the GDP of a superpower.

Interest in asteroid mining has been high enough that for a while there was something of an investment bubble. Companies like Planetary Resources and Deep Space Industries attracted tens of millions of dollars in investment capital in the early 2010s (much of it from heavy hitters in the technology industry) for their proposed asteroid mining projects. These companies didn't take off as hoped and wound up being acquired by other firms who seemed more interested in their technology than their asteroid-mining missions. (On the positive side, one reason they failed was that there were so many other attractive space companies offering a faster payoff to investors.) But newer companies, with perhaps more realistic missions and timelines, have entered the field such as the U.K.-based Asteroid Mining Corporation and the California-based TransAstra. In addition, exploratory probes are already visiting asteroids, in part to investigate their suitability for mining.

Maybe these new companies will make it; maybe they won't. Most of the pioneering private-launch companies of the 1980s failed, as is common in new industries – especially those involving cutting-edge technology and new markets. My own projection is that asteroid mining is more likely to succeed as part of an ambitious Earth-orbital construction plan building Solar Power Satellites, space hotels, or some sort of space manufacturing facilities, such that there's a ready-made market for the materials in space. Literally, anything in space is worth thousands of dollars per kilogram compared to lifting it from Earth. Even if that figure drops by a factor of ten or more, as Elon Musk promises, that's still a huge cost advantage. Materials delivered directly to Earth seem less promising. Even for commodities such as platinum, the earthbound price isn't that high. (Additionally, of course, an asteroid containing $6-$7 trillion of platinum at current prices wouldn't really be worth that much if delivered to Earth, because flooding the market with so much new supply would cause prices to plummet.)

Small-scale asteroid mining might involve extracting modest amounts of water and other volatiles from small carbonaceous asteroids to provide atmosphere and fuel for orbiting facilities. Any leftover rubble would still be valuable as insulation or radiation shielding. Remember, almost everything has value in space.

Moon mining is also on the horizon, and the U.S. government is showing considerable interest. There's convincing evidence that substantial quantities of water ice are available in shadowed craters on the poles, and that water could be easily obtained for use at lunar bases and refueling stations. Thanks to the Moon's low gravity, that water could be fairly easily transferred to facilities in Earth's orbit. More speculatively, lunar soil contains Helium-3 which is deposited by the solar wind. Although unavailable on Earth, Helium-3 is thought to be an idea fuel for fusion power plants, being much easier to fuse and producing less radioactivity in the process than fuels available on Earth. On the one hand, I've been hearing promises about fusion power my entire life; on the other hand, the physics surrounding it are well understood, and the presence of Helium-3 on the Moon is clear. The rest is "just engineering," as they say, although there's really no "just" about it.

We Fought the Law and the Future Won

One barrier to asteroid mining, as well as mining resources on the Moon and other celestial bodies such as comets, has now been breached. For many years, the legal status of space resource extraction was uncertain, which scared away many investors. But in 2020, President Trump issued an executive order that made things a lot clearer, though in an interesting bit of bipartisanship that order proceeded from a law signed by President Obama in 2015.

On April 6, 2020, President Trump issued an executive order entitled "Encouraging International Support for the Recovery and Use of Space Resources," that will likely jump-start the extraction of resources from both the Moon and asteroids to facilitate a permanent human presence in the solar system. In so doing, President Trump also ended an era of uncertainty and prevented the growth of a new, unaccountable international bureaucracy.

The 1967 Outer Space Treaty, which virtually

every nation signed, banned "national appropriation" and "sovereignty" over the Moon and other "celestial bodies" like Mars, asteroids, and comets, etc. It did not, however, ban the private appropriation of space resources. Although there are some who argue otherwise, it seems pretty clear that the Outer Space Treaty does not prohibit either private entities or governments from exploiting those resources, so long as they don't claim sovereignty over them, as Julian Ku, a law professor at Hofstra University, pointed out in *Opinio Juris* in late 2015. (Andrew Tingkang, an attorney, has argued convincingly that asteroids small enough to move cannot be considered "celestial bodies" under the Outer Space Treaty and thus are not covered by it.)

Since the Outer Space Treaty left space appropriation open to private industry, a group of mostly developing nations, acting under the auspices of the UN's "New International Economic Order" declaration of 1974, backed the 1979 Moon Treaty. That treaty banned any sort of private exploitation of space resources and mandated that any such activity only take place under the supervision of an international authority, with a rake-off, not surprisingly, going to developing nations. The Moon Treaty was a nonstarter in the United States, where space enthusiasts managed (with help from industry) to block ratification efforts, despite support from the Carter administration. The U.S. activists were not alone in their skepticism. As the White House noted in its statement accompanying the 2020 executive order, only seventeen of the ninety-five nations belonging to the UN's Committee on the Peaceful Uses of Outer Space ever ratified the Moon Treaty.

Nonetheless, many supporters of space development have feared that the Moon Treaty might be used by opponents to block development, or that political winds might shift. President Trump's new executive order makes clear that the United States does not regard the Moon Treaty as good law, and instructs the U.S. Department of State (which has at times been a bit squishy) to oppose any effort to regard the Moon Treaty as binding.

The executive order also orders the Department of State to work with "like-minded" nations to make bilateral and multilateral agreements that facilitate private resource extraction in outer space. This set of agreements, under the working title of the Artemis Accords, is designed to allow countries that have skin in the game – countries with active space capabilities and industries – to work out among themselves what needs to be done in order to move things forward safely and fairly. Though the Russians have grumbled, other nations have already expressed interest.

In some ways, this order is more evolutionary than revolutionary. In 2015, Congress passed and President Obama signed the U.S. Commercial Space Launch Competitiveness Act, which provided that: "A United States citizen engaged in commercial recovery of an asteroid resource or a space resource under this chapter shall be entitled to any asteroid resource or space resource obtained, including to possess, own, transport, use, and sell the asteroid resource or space resource obtained in accordance with applicable law, including the international obligations of the United States." This order ensures that the international obligations will be supportive, not destructive, of such efforts.

In another sense, this latest executive order is very in tune with the spirit of these times. In the forty-plus years since the Moon Treaty was drafted, international organizations have not exactly covered themselves with glory, with the recent failure of organizations such as the WHO and the E.U. to deal with the Wuhan coronavirus epidemic being the latest example. Nationalism is making something of a comeback, and this order reflects that. It's not isolationist since it encourages international agreements; but the approach is nation-to-nation, rather than being focused on international organizations and bureaucracies. This approach is likely to become the standard one in the coming years.

Things to Come

As you can see, we have all the ingredients for a new era in space: lower and falling costs; abundant resources and eager entrepreneurship; and all under the cover of a supportive government. What can we expect from here?

I write this in 2020, a year that has unsettled expectations even more so than the several years preceding it. But given the conditions described above, and assuming no global war or catastrophe, or a U.S. government that will politically crush the industry, I have a few ideas. (Even if the U.S. government did make such a mistake, many other countries ranging from China, which is rapidly building a commercial space industry in fairly open imitation of SpaceX, to Luxembourg, which wants to play the same role toward space companies that Panama plays to shipping and Delaware to corporations, will probably make sure that the revolution will still happen. It'll just be one that leaves the U.S. behind.)

The long-term goal looks a lot like what visionaries such as Konstantin Tsiolkovsky envisioned: substantial numbers of humans living off the surface of the earth. Both Jeff Bezos and Elon Musk see this as a goal and have talked of eventually moving most industry into orbit, leaving a much cleaner and healthier planet for the humans and other species who remain. (Writing in *The New Atlantis* recently, space expert Rand Simberg even called them "the new visionaries.")

There are differences. Elon Musk loudly favors Mars as a major site for human habitation. Bezos more quietly (as is his wont) favors the Moon. Both (but especially Bezos, who coincidentally is a Princeton alumnus) look favorably on a proposal by Professor Gerard K. O'Neill of Princeton University to establish orbiting habitats holding tens of thousands of people. (At an honors seminar at Princeton, O'Neill posed the question: is the earth's surface the best place for an expanding technological society? The answer he received was "no." This resulted in a very popular book on space colonies called *The High Frontier*, which has inspired many of our space entrepreneurs. The preferred location for such a colony, the stable Lagrange 5 libration point in the Earth/Moon system, gave its name to the L5 Society, an early pro-space group.)

The amount of energy and resources available throughout the solar system is, from our perspective, effectively limitless. A society no longer tied to the planet's surface could grow vastly larger (since the solar system could support many times Earth's population comfortably) and vastly more diverse. It would also be safer.

Right now, civilization on Earth is at risk from pandemics (natural or artificial in origin) far worse than the Wuhan coronavirus, volcanic eruptions, asteroid strikes (which disturbingly turn out to be more common than was thought just a few decades ago), and from other planetary catastrophes, including no doubt some as yet unknown and unimagined. (As astrophysicist Neil deGrasse Tyson observed after

The material returns from expanding into space are incalculably vast, but the spiritual ones may be just as great.

the Chelyabinsk meteor strike, "Asteroids are nature's way of asking: 'How's that space program coming along?'") A civilization that is not solely dependent on one planet, however, is by definition not susceptible to being wiped out by a planetary catastrophe. As Robert Heinlein once said, the earth is too small and fragile a basket to hold all of humanity's eggs.

An economy spanning first the Earth/Moon system and orbital space, and later extending to Mars, various asteroids, and eventually the solar system as a whole, would be vastly richer and much more resilient than anything that can exist on a single planet. It would also be more diverse socially, culturally, and intellectually, and that is something especially important now.

It is also something we should seek sooner rather than later. As the technologies that risk Earth's mass destruction – from nuclear to biological, and soon nanotechnology and artificial intelligence – continue to advance, the risk of remaining in a single basket grows. There is a window of vulnerability between the time that planetary civilization-wrecking weaponry becomes common, and the time that humanity is sufficiently dispersed to be safe from such a fate. At present, the trends are neatly summarized in Vernor Vinge's novel, *Rainbows End*.

> *Every year, the civilized world grew and the reach of lawlessness and poverty shrank. Many people thought that the world was becoming a safer place.* [. . .] *Nowadays Grand Terror technology was so cheap that cults and criminal gangs could acquire it.* [. . .] *In all innocence, the marvelous creativity of humankind continued to generate unintended consequences. There were a dozen research trends that could ultimately put world-killer weapons in the hands of anyone having a bad hair day.*

Written in 2006, Vinge's novel was set in 2025, and this observation seems to be pretty much spot-on, except perhaps for the shrinkage of lawlessness. For those thinking that a "go-slow" approach to space expansion is somehow more prudent or sensible, it's worth noting that each year of delay is another year of vulnerability. Sure, it might not matter if we're lucky. But it's 2020. Do you feel lucky?

Opening Up

Konstantin Tsiolkovsky wrote “the Earth is the cradle of the mind, but one cannot live in a cradle forever.” Walter McDougall, in the introduction to his excellent 1985 political history of the command-economy Space Age, *... the Heavens and the Earth: A Political History of the Space Age*, compares humankind’s emergence into outer space to the emergence of the first lungfish from the oceans. Though dramatic, this comparison is not overdrawn. The emergence of humanity into spaces beyond the earth represents a real qualitative change in human existence. Over the next few centuries, if things on Earth aren’t too catastrophically mismanaged, humanity will spread to a variety of locations throughout the solar system, carrying parts of Earth’s biosphere with it. Indeed, some writers have suggested that this is humanity’s real role. If one believes, as some proponents of the so-called *Gaia* hypothesis do, that the earth’s ecosystem can be treated as a single meta-organism, then humanity’s role may in large part be that of meta-gametes, carrying the seeds of life to new environments where it could not have evolved nor could it have reached in other ways.

Whether one accepts this description or regards it as so much science fiction, there can be no question that the expansion of humanity into outer space will be a drama and a challenge for which there is no real counterpart in human history. The great age of Earth exploration is the closest analog, but wherever European explorers went, there were people already there. By contrast, in exploring space, explorers really do go "where no man has gone before," to use the famous *Star Trek* phrase. While there may be space colonizers, there will be no space colonized because, unless we make a truly startling discovery, there will be no native inhabitants to exploit.

The benefits of this space expansion may not simply be material, but also spiritual. A society that is growing and offering fresh opportunities is likely to have a more positive mindset than one that is stagnant or shrinking.

For many years it was a staple of American historical thinking – and remains so in some quarters – that much of America's uniquely individualistic,

A society no longer tied to Earth's surface could grow vastly larger, since the solar system could support many times Earth's population.

opportunity-oriented character came from the centuries-long existence of the frontier. With a frontier, those dissatisfied with the status quo could set out to make a new life. Even those who stayed home (which was most people, after all) experienced life as more dynamic. The vigor of the East Coast's business community in the nineteenth century was, thus, as much a product of the frontier as it was of more local conditions.

This was Frederick Jackson Turner's famous "Frontier Thesis" from his book *The Significance of the Frontier in American History*. Although academic history has moved on to focus more narrowly on issues such as race, class, and gender, Turner's analysis has much to recommend it.

As Turner wrote,

> *To the frontier the American intellect owes its striking characteristics. That coarseness of strength combined with acuteness and inquisitiveness; that practical, inventive turn of mind, quick to find expedients; that masterful grasp of material things, lacking in the artistic but powerful to effect great ends; that restless, nervous energy; that dominant individualism, working for*

good and evil, and withal that buoyancy and exuberance that comes from freedom – these are the traits of the frontier, or traits called out elsewhere because of the existence of the frontier.

For a moment, at the frontier, the bonds of custom are broken and unrestraint is triumphant. There is no tabula rasa. The stubborn American environment is there with its imperious summons to accept its conditions; the inherited ways of doing things are also there; and yet, in spite of the environment and in spite of custom, each frontier did indeed furnish a new opportunity, a gate of escape from the bondage of the past; and freshness, and confidence, and scorn of older society, impatience of its restraints and its ideas, and indifference to its lesson, have accompanied the frontier. [...] *What the Mediterranean Sea was to the Greeks, breaking the bonds of custom, offering new experiences, calling out new institutions and activities, that and more the ever-retreating frontier has been to the United States.*

The American frontier has been gone for more than a century, but the idea remained strong enough that John F. Kennedy, who launched America to the Moon, dubbed his campaign theme "New Frontiers," a phrase that still resonates. In *The Case for Mars*, author Robert Zubrin cites Turner's passage above and observes:

> *Perhaps the question was premature in Turner's time, but not now. Currently, we see around us an ever more apparent loss of vigor of our society: increasing fixity of the power structure and bureaucratization of all levels of life; impotence of political institutions to carry off great projects; the proliferation of regulations affecting all aspects of public, private, and commercial life; the spread of irrationalism; the banalization of popular culture; the loss of willingness by individuals to take risks, to fend for themselves or think for themselves; economic stagnation and decline; the deceleration of the rate of technological innovation. Everywhere you look, the writing is on the wall.*
>
> *Without a frontier from which to breathe new life, the spirit that gave rise to the progressive, humanistic culture that America has represented for the past two centuries is fading. The issue is not just one of national loss – human progress needs a vanguard, and no replacement is in sight.*

I agree. I would rather live in a society possessing the characteristics Turner describes than the one that Zubrin, all too accurately, reports. Unfortunately, in the twenty-first century we have achieved the "global village" discussed in the twentieth century, but we are also rediscovering how intellectually oppressive, conformist, enervating, and stultifying village life can be.

There is an even older example. In his classic biography, *Admiral of the Ocean Sea*, Samuel Eliot Morrison describes the impact of Columbus's voyages of discovery on a continent sunk in stagnation and despair:

> *At the end of 1492 most men in Western Europe felt exceedingly gloomy about the future. Christian civilization appeared to be shrinking in area and dividing into hostile units as its sphere contracted. For over a century there had been no important advance in natural science and registration in the universities dwindled as the instruction they offered became increasingly jejune and lifeless. Institutions were decaying, well-meaning people were growing cynical or desperate, and many intelligent men, for want of something better to do, were endeavoring to escape the present through studying the pagan past.* [...]

Yet, even as the chroniclers of Nuremberg were correcting their proofs from Koberger's press, a Spanish caravel named Nina scudded before a winter gale into Lisbon with news of a discovery that was to give old Europe another chance. In a few years we find the mental picture completely changed. Strong monarchs are stamping out privy conspiracy and rebellion; the Church, purged and chastened by the Protestant Reformation, puts her house in order; new ideas flare up throughout Italy, France, Germany and the northern nations; faith in God revives and the human spirit is renewed. The change is complete and startling: "A new envisagement of the world has begun, and men are no longer sighing after the imaginary golden age that lay in the distant past, but speculating as to the golden age that might possibly lie in the oncoming future."

Christopher Columbus belonged to an age that was past, yet he became the sign and symbol of this new age of hope, glory and accomplishment. His medieval faith impelled him to a modern solution: Expansion.

Space is full of energy that you don't need to do anything on Earth to produce.

Today, of course, Columbus is even less fashionable than Turner, but that hardly defeats the point. Even Columbus's critics would generally acknowledge that his discoveries were good for Europe; their objection is to the impact on indigenous peoples. In space, on the other hand, at least in our solar system, there are no indigenous people to worry about.

It is perhaps no coincidence that the period some call the "golden quarter-century," from 1946 to 1971, was one of enormous optimism and growth, or that the decade following America's retreat from human space exploration was one of shrinking horizons and focus on "limits to growth," limits that, as it turned out, were far more psychological than real. The material returns from expanding into space are incalculably vast, but the spiritual ones – including the boost to the "animal spirits" of the business and cultural worlds – may be just as great.

Next Steps

In previous pages, I've laid out the three phases of space exploration and development to date: the visionary phase of Tsiolkovsky, Goddard and von Braun; the command-economy phase of Apollo and the Space Shuttle; and the now-blooming sustainability phase in which space activity will be, for the most part, economically self-supporting. Each of these phases is, and has been, important. Without the visionary phase, the command-economy phase might never have taken place; or if it had, it would have been short of ideas and basic concepts. The command-economy phase, meanwhile, proved that the visionaries' ideas were not impossible, but entirely possible, if perhaps somewhat expensive. The sustainability phase, meanwhile, will make the visionaries' ideas affordable.

Nothing is certain, of course. An unfortunate change in global events could snuff the current revolution out. On the other hand, good policy could accelerate things considerably.

We are fortunate enough to have good policy under the current administration, and for that matter under the previous Obama administration. Good policy for the commercial space revolution mostly involves leaving things alone, with perhaps a few policy tweaks designed to promote innovation. That will remain true in the coming decade. But there are a few other things that can be done.

In the 1920s, the United States Post Office subsidized airmail carriage in a fashion intended to speed the development of long-range, reliable, and affordable air transportation. This policy, which rewarded success but did not subject aircraft manufacturers to heavy-handed government control, worked very well. In 1987, law professor Robert Merges and I argued for a similar approach to space industries in an article entitled "Toward an Industrial Policy for Outer Space" that was published in *Jurimetrics*, the American Bar Association's law, science, and technology journal. In more recent years, NASA's purchases of commercial services such as supply launches to the International Space Station have followed a model that somewhat resembles what we had recommended.

This approach should be expanded and intensified. The United States government shouldn't be buying or building spaceships. It should be purchasing launch services from people who do. (Trying to centralize all spaceflight into a single government-built spacecraft, the Space Shuttle, was a disaster, and we should basically try to do the exact opposite.) The more the U.S. acts like a commercial customer, the more it will encourage companies to produce things that have value in the commercial markets. Generally, this will be a better deal for taxpayers, especially over the longer term as costs go down.

In addition, regulation should be kept to the necessary minimum (we're already doing a pretty good job at that) and barriers to entry should be kept as low as possible. Competition is good.

For more ambitious projects, prizes are an effective and relatively inexpensive way to encourage interest. When Charles Lindbergh flew the first nonstop transatlantic flight less than a century ago, he did so in response to the Orteig prize which was offered for just such an accomplishment. Had Lindbergh not made it, one of the numerous other competitors for the prize would have. That's one of the advantages of a prize. They tend to attract far more investment from competitors than the prize itself represents.

More recently we've seen how successful prizes can be. The Ansari X Prize, awarded for the first launch of a human-bearing spacecraft into space and its safe return to Earth followed by a repeat effort within two weeks with the same craft, was intended to promote reusable spacecraft and succeeded. The winner was Burt Rutan's SpaceShipOne. Its successor, SpaceShipTwo, is now flying for Virgin Galactic and will soon take paying passengers into space.

The Google Lunar X Prize, funded by Google founders Larry Page and Sergey Brin, offered $20 million to the contestant who landed a probe on the Moon that covered at least 500 meters and transmitted high-definition video and images back to Earth. There were other prizes for steps along the way. Results were mixed. No one was able to claim the prize, which may have been a bit too ambitious, but from the standpoint of attracting investment in space technology it was a success. Several milestone prizes were awarded, and according to the X Prize Foundation, it attracted more than $300 million in investment and led to the creation of new space-related companies in several countries. Not bad for a prize that was never actually awarded.

Of course, one prize that encourages investment is a property right in the result. The United States's movement to recognize property rights in space, through the 2015 bill that President Obama signed and the 2020 executive order from President Trump, represents perhaps the biggest prize of all: a guarantee from the most powerful nation on Earth that discoverers will be able to enjoy the fruits of their efforts.

Conclusion

Having passed through the visionary and command-economy phases of space development, we are now entering the economically sustainable phase. The good news about this phase is that we are no longer dependent upon the whims of politicians. Progress today can proceed at its own pace, with the only sales pitch needed being the one aimed at paying customers. In essence, all we need to do to make progress now is to avoid screwing things up.

Will we manage that? I hope so, though it's not guaranteed. There are certain people who are offended by the very idea of humanity moving out into space, as if the concept of open horizons is inherently offensive. Maybe to them it is. Fortunately, their numbers seem to be pretty small at present. But it will be important to avoid and, if necessary, frustrate political efforts to shut space ventures down. Just because you're not dependent on the government for funds doesn't mean it can't put you out of business.

Of course, if the United States doesn't move forward in the space arena, others will – most likely China, which has a strong interest in space for military and civilian reasons, and which is looking for ways to gain prestige at the expense of longer-established powers. But culture matters. Latin America, first colonized by Spanish and Portuguese settlers, would look much different had it first been settled by the British or the Germans. Space will look much different, even centuries later, if it is first settled by the Chinese as opposed to the Americans. I would prefer the latter.

We've seen enormous results that make it possible to do great things in space without depending on political whims.

At any rate, I hope that this short treatment has provided a useful idea of where we came from, how we arrived at this point, and what we're doing now in what is perhaps the single most important aspect of human endeavor in the early twenty-first century. All we need to do to flourish is to keep moving forward. Faster, please!

First American edition published in 2020 by Encounter Books, an activity of Encounter for Culture and Education, Inc., a nonprofit, tax exempt corporation.
Encounter Books website address: www.encounterbooks.com

Manufactured in the United States and printed on acid-free paper. The paper used in this publication meets the minimum requirements of ANSI/NISO Z39.48–1992 (R 1997) (*Permanence of Paper*).

FIRST AMERICAN EDITION

LIBRARY OF CONGRESS CATALOGING-IN-PUBLICATION DATA

Names: Reynolds, Glenn H., author.
Title: America's new destiny in space / Glenn Harlan Reynolds.
Description: First American edition. | New York : Encounter Books, 2020. | Series: Encounter intelligence | Includes bibliographical references. |
Identifiers: LCCN 2020029549 (print) | LCCN 2020029550 (ebook) | ISBN 9781641771825 (paperback) | ISBN 9781641771832 (epub)
Subjects: LCSH: Astronautics—United States. | Outer space—Civilian use.
Classification: LCC TL789.8.U5 R47 2020 (print) | LCC TL789.8.U5 (ebook) | DDC 338.4/762940973—dc23
LC record available at https://lccn.loc.gov/2020029549
LC ebook record available at https://lccn.loc.gov/2020029550

10 9 8 7 6 5 4 3 2 1